Story Selling in the Connected Economy

Build Trust and Retain Customers for Life

by

Bill Whitley & Patrick Thean

authorHOUSE

1663 LIBERTY DRIVE, SUITE 200
BLOOMINGTON, INDIANA 47403
(800) 839-8640
www.authorhouse.com

© 2004 Bill Whitley & Patrick Thean
All Rights Reserved.

No part of this book may be reproduced, stored in a retrieval system, or transmitted by any means without the written permission of the author.

First published by AuthorHouse 05/24/04

ISBN: 1-4184-5593-8 (e)
ISBN: 1-4184-5108-8 (sc)
ISBN: 1-4184-7071-6 (dj)

Library of Congress Control Number: 2004105310

Printed in the United States of America
Bloomington, Indiana

This book is printed on acid-free paper.

Acknowledgements

The success of any writing project depends on the efforts and experiences of many people whose names never appear on the cover. We, the authors, are just the fortunate ones who were able to record and express our journey.

We would like to thank:

The MindBlazer team and our clients who provided the energy and experience for this book. Amy Wellman who edited the final version.

Most of all, we are eternally grateful to our families. Their contributions to our lives, experiences, and this book are far greater than they would imagine. Thank you to Lee-Anne, Pei-Yee, our parents and our children for their inspiration, support and love.

Table of Contents

INTRODUCTION 5

Part I

The Problem

CHAPTER 1
Trends.. 11

Part II

The Solution

CHAPTER 2
Education-Based Marketing..21

Part III

The Process

CHAPTER 3
A 5-Step Process 35

CHAPTER 4
Step 1 – Observe 41

CHAPTER 5
Step 2 – Visualize... 49

CHAPTER 6
Step 3 – Design 67

CHAPTER 7
Step 4 - Develop... 81

CHAPTER 8
Step 5 - Rollout 91

Part IV

The Technology

CHAPTER 9
Technology of Story Telling 107

CHAPTER 10
Lead Capture 123

NOW IT'S YOUR TURN 135

GLOSSARY 137

INDEX 139

Introduction

"I may not have gone where I intended to go, but I think I have ended up where I intended to be."

-Douglas Adams

"I do not think much of a man who is not wiser today than he was yesterday."

-Abraham Lincoln

"Everybody wants to be somebody; nobody wants to grow."

-Johann von Goethe

Our Strange and Wonderful Journey

Several years ago, my partner Patrick Thean and I formed a company called MindBlazer. Our vision was (and is) that the TV world and the Internet world are converging. Within our industry, this converged space is being called everything from *new media* to *rich media* to *streaming media* to *web casting*. To be honest, I don't really care what this space is ultimately called. What I'm most excited about is that this convergence brings:

- New ways to educate customers

- New ways to attract prospects

- New ways to train employees.

Patrick and I personally are a good example of the convergence between technology and TV. Patrick is a technology guru who created and sold a transportation logistics software company called Metasys. I started an interactive multimedia company called The Whitley Group and later sold to one of the nation's largest strategic Internet consulting firms.

Together, we created MindBlazer to help corporations harness the power of this converged space. I focus on helping clients create valuable content and Patrick focuses on delivering the content with technology. When we wrote this book, I focused on the chapters that deal with content and Patrick wrote those that deal with technology.

Over the last four years, we've helped a variety of large corporations create education-based marketing programs on video, CD-ROM and other online and offline channels. Our clients have included American Express, Cigna, Cisco Systems, LendingTree, Lowe's, State Farm, Sysco Foods, Wachovia and Yahoo!. What we quickly realized is that, although corporations are interested in participating in the brave new world of broadcasting their content on-demand over the Internet, they face a huge challenge. Content that works well in the brick-and-mortar world can be amazingly dull in the broadcast world.

When it comes time to do a webcast, many corporate communicators aim a camera at a subject matter expert and shoot "talking head" video.

Unfortunately, many failed webcasts later, communicators wonder why their audience didn't tune in to watch their content.

Here's why: over the past 50 years, the TV world has honed its story telling prowess and has become expert at creating content that attracts and retains audiences. TV content is much more than "talking head" video. Informational TV content is built around "story packages." Usually the term implies a reporter has interviewed several people, added a voiceover, and edited the footage into a "package." Businesses must keep this concept in mind when producing informational TV.

In a typical package you will meet a variety of people with different points of view. We call them *heroes, luminaries and experts*. This makes the content more interesting and more credible than having one corporate presenter. Story packages last about one minute and 42 seconds. They're designed to be light, easy to understand and very applicable to the viewer. When was the last time you saw a corporate presentation that was light, easy to understand and applicable to the viewer? If your experience is similar to mine, the answer is never.

What we've learned is that to be effective, corporations needed help creating content that could attract and hold viewers. To accomplish this, we formulated a five-step process for developing high-value content (more on that later). A critical turning point for MindBlazer came when Yahoo! Broadcast heard about this process. A few months later, we became their business partner. With the acquisition of Broadcast.com, Yahoo! had quickly become the

800-pound gorilla in the streaming media world. Live, on-demand, mock live, you name it, Yahoo! Broadcast had it.

When Yahoo! sold a major marketing webcast, MindBlazer was brought in to design and produce the content. Over two years, we produced over 50 video-on-demand, marketing webcasts for Yahoo! customers. This was a wonderful period for MindBlazer. With 30 Yahoo! reps selling our services, we grew quickly. We helped Yahoo! close deals and they introduced us to wonderful customers.

What we learned from these experiences is fascinating. We now know that if you provide someone with helpful, friendly advice, you can build trust. If you focus more on solving their pain than selling your product, you can become their trusted advisor. And if you diligently follow a process, you can consistently create wonderful content.

We have designed this book as a step-by-step approach to doing exactly that… to create a road map that will help you capture what your business knows and package that knowledge into stories. These stories will build your customers' and prospects' trust in you. When you are their trusted advisor, your company's potential is limitless. Good luck. We look forward to hearing your results. You can email either Patrick or me at:

bill@mindblazer.com

patrick@mindblazer.com

Part I

The Problem

CHAPTER 1

Trends

- **SALES & MARKETING IS MORE CHALLENGING THAN EVER BEFORE**

TV & The Internet Are Converging

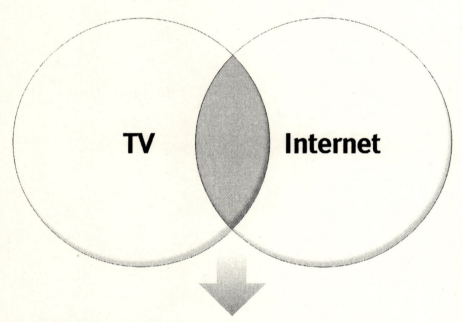

(New Media, Rich Media, Streaming Media, Webcasting)

Trends

"People only see what they are prepared to see."
-Ralph Waldo Emerson

"Choice has always been a privilege of those who could afford to pay for it."
-Ellen Frankfort

"America is not only big and rich, it is mysterious…"
-David Riesman
The Lonely Crowd

Sales & marketing is more challenging than ever before

In the good old days there were three TV stations, three car makers and three brands of beer. Life was easy for marketers. If you wanted to reach America, you could buy advertising slots on three networks and a couple of magazines and you were done. Now, sales and marketing is more challenging. As advertisers bombard consumers, consumers are screening out and ignoring most messages.

An increase in information

Thanks to the Internet, consumers have more information about what they are considering buying than ever before. BMW, for instance, recognizes that 85% of their buyers research the purchase on the Internet prior to making the purchase. If you type in "BMW" and "Consumer Information" on Google, you get over 3,000 sites dripping with information. Some sites like www.epinions.com allow you to read third-party reviews and see how BMW owners rated their cars. Other sites like www.consumerguide.com rate cars on five categories with scores for each. And of course sites like www.autofinder.com allow you to search for new and used cars, find the best price and buy. You can even find out what a dealer paid for a car and base your offer on that information.

This change isn't true just for cars. People are turning to the Internet for insight on all kinds of purchases. Consumers today have more information, and that information gives them more power.

An increase in choices

Consumers also have more choices than ever in almost every category, from TVs to cars, cereal to beer, and jewelry to cell phones. This makes the marketer's job tougher. When you offer one of a hundred choices, how do you differentiate yourself?

An increase in marketing messages

And finally, consumers are overloaded with sales messages. Advertising and marketing channels have proliferated. It used to just be TV, radio and print ads. Now we have banner ads, pop-up ads, direct mail and telemarketing. Ads are even showing up on exercise bikes, and in elevators and taxicabs. This information overload is causing consumers to screen out messages.

The question then is: How do you win? How do you succeed in this more challenging world of selling and marketing to an empowered consumer? In a word, the answer is trust. More specifically, focus on building trust. In a crowded, noisy market, relationships are more powerful now than ever before. As products get more sophisticated and product choices multiply, consumers are looking for someone to trust. Consumers want helpful, friendly advice that helps them make the best decision. In the next chapter you will learn how to develop trust and create significant relationships. You will learn about a powerful technique called education-based marketing.

Chapter Summary:

Bottom Line: Sales & Marketing is More Challenging

- Thanks to the internet, consumers have more information than ever before.

- Consumers have more choices than ever before.

- Consumers have been bombarded with marketing messages for so long that they now filter out most of the messages.

Bottom Line: Information and Power

Consumers are now armed with information and power. The internet has allowed consumers to selectively educate themselves on what they need before purchasing. This information is no longer controlled by the provider of products and services. Consumers have access to third-party advice and various consumer guides to help them make informed decisions before buying products.

How?

Fight fire with fire. Give your customers helpful and friendly advice that helps them make the best decisions.

Thought Starters

1. What trends have you seen recently that upon reflection you wish that you have acted and taken advantage of?

2. How did neglecting the trend that you previously noticed impact your business?

Part II

The Solution

CHAPTER 2

Education-Based Marketing

- **An early lesson in education-based marketing**
- **The worry of sharing too much**
- **Making it work**
- **Multiple delivery channels**

Knowledge Capture & Packaging

Industry Knowledge
- Heroes (Customers)
- Luminaries (3rd Party)
- Corporate Expert

Customers / Prospects
- Face a variety of challenges
- Looking for solutions
- Want trusted advisors

Knowledge Packaging
- Story Packages
- Transferrable Lessons
- Strong Attracting & Holding Power

- Web
- CD/DVD
- Verbal
- Print

Education-Based Marketing

"It is one of the most beautiful compensations of life, that no man can sincerely try to help another without helping himself."

-Ralph Waldo Emerson

"Every problem has a gift for you in its hands."

-Richard Bach

"People seldom refuse help, if one offers it in the right way."

- A. C. Benson

An Early Lesson in Education-Based Marketing

I'll never forget one of my family's visits to my grandmother's house. I was a senior in college. My classes included an investment course on portfolio management. My grandmother, like any good grandmother, asked, "So Bill, what are you learning?" I told her what I'd learned about investing: "People like me, at younger years in life, are interested in growing our wealth. We're willing to take more risks in order to get higher returns." I went on to say, "People like you, who have retired, are much more interested in safety, liquidity and income. At your age, asset growth is no longer a priority. You want to protect and preserve what you have so that you can maintain your retirement lifestyle."

After I delivered this little speech, my grandmother looked over at my dad. (By the way my dad's name is Jim, he's a stockbroker and he manages my Grandmother's money). My grandmother said, "Jim, I would put my money with this boy." It was a wonderful feeling. I had just discovered the power of education-based marketing, or EBM for short. Looking back, I now realize that what I did can easily be done for a wide variety of topics. If you can articulate someone's problem and then share helpful, friendly advice, you can build instant credibility. You will position yourself as the kind of person that people want to do business with. You will become the trusted advisor.

Unfortunately, we all have to sell to people other than our grandmothers. Here's how this helpful, friendly style of selling may be delivered on a larger scale. Let's suppose I was a financial advisor and wanted to grow my practice. One good way to raise awareness and build credibility would be to put on a marketing seminar. If I were a financial advisor, the name of my seminar would be, "The top three ways to grow and preserve your wealth." If I presented it properly (which is a big if, because most people botch this step), I would share lots of good ideas and helpful information with my audience. This practice is known as abundant sharing. The entire seminar would come across as very much of a soft-sell approach. If you do that kind of seminar properly, in almost every case your audience will think: "This guy is smart. I trust this guy. When it's time to go forward, I think I'll do it with this person."

Here's a good example of the power of becoming a trusted advisor. A major food services company was trying to improve the image of its sales staff. The company wanted its reps to be seen as value-added consultants, not just order takers. One of their stars in this effort was the rep who called on a 10-location restaurant chain.

To add value, the company offered customers a free menu analysis and a menu design. When the rep analyzed the menu for the chain, he realized that two fast moving items, coffee and fried chicken, were under priced. He recommended the restaurant chain increase the price on both items.

The result? The restaurant made $500,000 in profit. This restaurant chain now views their rep as a valuable resource and will always buy from his company.

The worry of sharing too much

Perhaps you're still not convinced. You might worry if you share too much information, prospects won't have a reason to buy from you. You wonder if they'll simply take your knowledge and implement it themselves. I have found the opposite is true: the more you give away, the more trust you build. Here are a few reasons why:

- I believe the execution of an idea is much more important than the idea itself. Think about how many great products and services you've heard about that failed because of poor execution.

- Others don't have the expertise and time to act upon your ideas. That's why they turned to you. Especially if you're asking them to spend a lot of money, you have to demonstrate you know what you're talking about and their investment in you will be worthwhile.

- When you share your expertise, you build credibility and give the shopper confidence you can do what you claim. With their increased confidence comes perception of lower risk. If prospects trust you, they believe it's a smaller risk to choose you rather than hire someone else or do it themselves.

- By sharing your expertise, you tap into an important trend. Consumers are changing the game. They are taking back their buying power. They want relevant advice to solve their problems. This form of education-based marketing adds relevance to your brand by providing advice to solve problems. As marketers, we must provide relevant advice on topics that consumers care about – or our efforts are a waste of time.

Making it work

Topics that work

This sort of marketing works best for subjects I think of as "big, sweaty and passionate." In other words, big-ticket items, things you sweat and worry over, or things you might be passionate about. They naturally lend themselves to pre-purchase research. If you have a "big, sweaty, passionate" product, helpful, friendly advice is always appreciated. Select a topic that involves at least two of these critical elements: expense, worry, and passion. Having all three is even better.

Renovating your kitchen is a good example. Kitchens often cost $15,000 or more to renovate. They are difficult to design and provoke a lot of worry about whether it will be done right. And people feel passionate about their kitchens because they spend so much time there. Once you have a great topic, the next step is to capture and package high-quality stories.

Tread new ground

I've also found this style of marketing can be much more creative than the usual tactics. Consider a recent fund-raising effort by The McColl Center for Visual Arts, a non-profit community where artists who are accepted into the program can perfect their skills for free.

The traditional way a non-profit raises funds is to tell the audience what the organization does, why it's so wonderful, and ask for money. The McColl Center embraced education-based marketing and tried a fresh strategy. Center staff invited 450 people to come to a luncheon. At the luncheon, they showed a custom video presentation on two life-changing lessons that we can learn from artists. The first is how to live a life of passion. Most people are content to follow a career they don't love, if they make enough money. Artists, on the other hand, follow their dream above all else. The second lesson is the joy that comes from self-expression. Most artists will tell you the reason they create is a strong need to express themselves. The audience learned that it's possible to express yourself in the business world and feel the same joy.

That fund-raiser was extraordinarily successful – all by trying something new.

Reach 'em when they want to be reached

We have to reach consumers at the time and place of their choosing, using the mediums that they are already using. So a campaign strategy must employ multiple channels and mediums to reach various consumers in the target audience. Education-based marketing should be a multi-channel approach, deployed in these ways:

Multiple delivery channels

On-demand over the Internet

Take advantage of the Internet to extend your reach across geographic boundaries and time zones. This is a great way to deploy your marketing program because of the tremendous distribution available to you. This method is also very measurable and trackable. You will be able to track how many viewers have watched your programs and for how long. As more and more people commit to having broadband Internet, this becomes an even more powerful way to reach them. However, it is important to know your audience. One of our clients pointed out that only 20% of their customers were online, so they needed other offline channels to reach the rest.

CD-ROM/DVD

CD-ROMs and DVDs are very flexible channels. The cost of putting content on a CD-ROM or DVD is inexpensive compared to print media, especially if you require a large number of discs made. These discs can be used in direct mail campaigns, given to customers at store locations, or inserted in print advertisements in magazines.

Print

Key points of your content can also be condensed and summarized into single-page print materials that can be used to reach more traditional customers who still prefer the look and feel of paper and the written word.

Point of Sale Locations

We refer to this as education-based sales. Store sales staff can be taught your education-based marketing message and deliver a consistent, trust-building message to your customers.

Edumercials on cable TV

Edumercials are very similar to infomercials. The difference is that an edumercial would focus on solving the consumer's problem, while infomercials tend to focus on product features and benefits.

No matter the marketing channels you deploy, this point can't be emphasized enough: When you share helpful, big-picture advice, you tend to be viewed as the trusted advisor. You become the kind of person that people want to do business with.

Is it possible for anyone to become a trusted advisor? In most cases the answer is yes. In Part Three you'll learn a step-by-step process for capturing knowledge and packaging it into stories and lessons that will position you or your firm as the trusted advisor.

Chapter Summary

Bottom Line

Don't worry about sharing too much; you're more likely to be perceived as the trusted advisor than as unnecessary.

EBM is a wholesome approach to marketing that provides more value than the typical product-centric sales pitch.

For maximum impact, use multiple channels to reach your audience at the time and place of their choosing.

How?

Share helpful, big picture advice, and you will be viewed as the leader in your industry.

Thought Starters

1. What advice can you give your customers that will build trust instead of selling product?

2. Can you recall an instance where helpful advice moved you to buying a product from someone? Did you continue to buy other products from them?

Part III

The Process

CHAPTER 3

A 5-Step Process

- **How to capture knowledge and package it into stories and lessons that increase sales**
- **Why it's important to follow a process**
- **Key benefits**
- **Avoid the common pitfall**

The MindBlazer 5-Step Creative Process

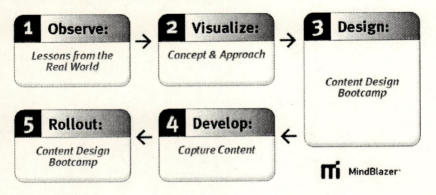

A 5-Step Process

"It is good to have an end to journey toward, but it is the journey that matters in the end."

-Ursula K. LeGuin

"A journey of a thousand miles, begins with a single step"

-Unknown

"Not all who wander are lost."

-J.R.R. Tolkien

How to Capture Knowledge and Package it into Stories and Lessons that Increase Sales

The process of capturing knowledge and packaging it into stories and lessons is easier than you may think. We will introduce you to a five-step process you can follow to consistently capture and package high-quality stories.

Why it's important to follow a process

Anytime you tackle a complex project, it pays to have a step by step process. A process will keep the team on track and enable clear communication with all team members. A published process also raises your sponsor's confidence for a successful project.

Key Benefits

I wanted to share with you first hand how we have benefited from having a consistent process. Over the last three years, we have delivered more than forty projects to over thirty customers. Here are some benefits that we have reaped over the years:

- Sponsors have been pleasantly surprised that we were able to harness creativity with a step by step process. Customers who purchased multiple projects from us had a consistent customer experience from project to project.

- We have used the QA steps in our process as quality checkpoints. We developed checklists around the QA checkpoints to make sure that we have done everything we need to at each major phase of our projects.

- Our process has become a basis for continuous improvement and learning. We are able to review project deliverables and compare them to what our process tells us we should achieve. We have continued to improve our process.

- We have been able to pass on project experience to new employees and maintain our consistent delivery with new employees.

- It has provided us with a language that we are all familiar with and improved communication in our project teams.

Avoid the common pitfall

Your process should be a living process, not a dead one.

Many people fall into the trap of using their process to dictate every nuance of a project in a way that substitutes the brain and independent thinking with *the process*. This leads to bureaucratic and inefficient teams. Never discourage your team to stop thinking just because you have a process. Instead, use your process as guiding posts for your project. Allow new ideas and thinking to improve your process. Then allow your process to be the vehicle to disperse that newly acquired knowledge to the rest of your team.

Chapter Summary

Bottom Line:

A process or method gives you a consistent way to deliver projects to both internal and external sponsors. It allows a clear way to communicate project progress and ensure project quality through well defined quality assurance milestones.

How:

Document your process. Make your process a living process, providing a platform to receive feedback and improve your process as your team gathers more experience.

Thought Starters

1. Do you have a consistent method for passing project knowledge on to new employees?
2. Have you noticed any successful patterns in delivering results to your clients? Were you able to repeat that success with other clients?

CHAPTER 4

Step 1 – Observe

- **OBSERVE THE REAL WORLD AND IDENTIFY WHAT YOUR AUDIENCE WANTS TO LEARN**

The MindBlazer 5-Step Creative Process

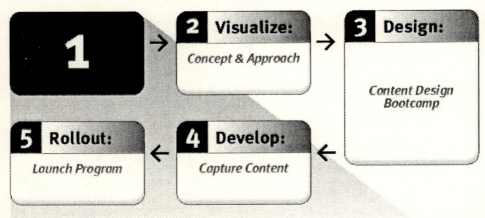

1 Observe: Lessons from the Real World

1. Observe the Real World

- *Audience (User & Power User)*
- *Sales/Customer Service*
- *Luminaries & Authorities*
- *Publications*
- *Tangents*

2. Identify Themes

Observe

"In the fields of observation chance favors only the prepared mind."

-Louis Pasteur

"To linger in the observation of things other than the self implies a profound conviction of their worth."

-Charles-Damian Boulogne
My Friends the Senses

"There is no more difficult art to acquire than the art of observation, and for some men it is quite as difficult to record an observation in brief and plain language."

-William Osler, Aphorisms from
His Bedside Teachings and Writings

Observe the real world and identify what your audience wants to learn.

One of the biggest challenges that corporate presenters have is that in most cases their presentations are too vendor-centric. Many times the presentations and demonstrations are beautifully done, with graphics, animations and even video. But they usually fall short in one major category… they focus too much on features and benefits. The story is told only from the vendor's point of view. They tend to "pound their chest" with messages saying, "we do this" and "we do that". Remember, as a consumer I'm not interested in your product. I'm interested in solving my problem. When clients buy from MindBlazer, I know they aren't buying an education-based marketing program per se. They are buying an opportunity for increased profits. That's really what they want.

All consumers have needs and reasons why they buy. Here are some of them:

- to save time
- to save money
- to make money
- to feel safer
- to be more attractive
- to feel better about themselves
- to appear more successful
- to provide for their future
- to simplify or make their lives better
- to give themselves pleasure
- to care for their family members.

The first step to tapping into these needs is to identify your prospect's biggest challenges. The best way to do that is to observe the real world. The best way to "observe the real world" is to talk to a variety of people who come in contact with your product or service. Start with your best sales reps or customer service people. They can give you a quick overview of who the typical customer is, what the customer's biggest challenges are, and how your firm solves their challenges. But don't stop there. Talk to a few customers. Ideally you will want to talk with both *typical users* and *power users*. Typical users can tell you what their main challenge is and how they solved it. A power user can give you even more. For example, let's say you wanted to gather advice about "how to keep your home germ- and disease-free." A new mom may have taken measures to create a safe, germ-free environment for her newborn. She can give you her perspective. A power user might be someone who runs a daycare, who is under government regulation to keep a very clean environment with a bunch of kids underfoot. In addition to personal challenges, the power user can typically talk about all the challenges a typical customer may face and the best ways to solve them. These real world lessons can be invaluable when you are crafting your story.

The next source of real world wisdom comes from *luminaries* and *authorities*. Luminaries and authorities are third-party experts on your topic. In our "How to Keep Your Home Germ Free" example, a good luminary might be someone who writes articles for *Good Housekeeping* or *Parents*. The ultimate authority on the topic may be someone from the Centers for Disease Control.

One additional source of real world wisdom comes from a source we loosely define as *tangents*. Tangents are involved with the category, but not directly.

For example, if I wanted to do research on how to design the ideal landscape plan, a Feng Shui expert may be able to give me good advice on landscaping from a Feng Shui perspective.

I suggest conducting a total of at least 10 to 12 interviews from at least four different groups. Take copious notes. As the wise old saying goes, "The palest ink is stronger than the best memory." During the interview process you should start to uncover some consistent themes. You will begin to notice that several of the challenges seem to get mentioned more than others. These challenges will become your themes. These themes will become the foundation of your story. If you answer the question of how to solve these challenges, you will position yourself as the trusted advisor. You won't be "pounding your chest." You will be creating content the viewer genuinely wants to see.

Chapter Summary

Bottom Line

Step 1: Observe the real world.

Seek to understand your prospect's biggest challenges before building content.

How?

Get a real first hand understanding of the prospect's needs and challenges by observing the real world. Observe customers and prospects, talk to sales and customer service staff and industry luminaries, and read publications. Learn about what your target consumers really care about. Once this is accomplished, you can now build content to address these needs and challenges. This approach will position you as the trusted advisor on these topics.

Thought Starters

1. Think of an example where you overwhelmed a prospect with product features and benefits thereby losing the sale. Would the outcome have been different if you had offered helpful and friendly advice first? How would you do this differently today?

CHAPTER 5

Step 2 – Visualize

- **VISUALIZE THE CONCEPT AND THE APPROACH**
- **THE BODY**
- **THE TEASE**
- **THE TRANSFERRABLE LESSON**
- **GAINING EXECUTIVE BUY IN**

The MindBlazer 5-Step Creative Process

2 Visualize: Concept & Approach

1. Brainstorm Ideas Based on Themes

2. Develop Concept

- *Content Outline*
- *Use Case Scenario*
- *Business Case*

Visualize

"In creating, the only hard thing is to begin: a grass blade's no easier to make than an oak."
-James Russell Lowell

"Joy is but the sign that creative emotion is fulfilling its purpose."
-Charles Du Bos
What Is Literature?

"The worth of a book is to be measured by what you can carry away from it."
-James Bryce

<u>Visualize the concept and the approach</u>

Once you understand the true challenges your customer is trying solve, you can begin crafting a story that demonstrates how you solve those challenges. There are three pieces of the story that should be developed:

- The Body
- The Tease
- The Transferrable Lesson.

The Body

There are four main ways to organize the body of your story:

- Sequential
- Category
- Bold Comparison
- Problem/Solution.

Sequential

If your story unfolds over time, the sequential format may work well. As its name implies, a sequential story tracks the sequence of key events. For example, one of the great American product innovation stories is the tale of how the Post-it® Note was created. Here's how the Post-it® Note story would be told using the sequential format:

> In 1968 Dr. Spence Silver, a research scientist for 3M, came up with an unusual adhesive. The new adhesive did not stick very strongly. It was interesting, it was different, and it was useless. Or so it was thought at the time.
>
> Twelve years later, Art Fry, another 3M product development researcher, was having trouble keeping bookmarks in place while singing in his church choir. Fry took some of Dr. Silver's adhesive and applied it along the edge of his bookmark. His church hymnal problem was solved!

Fry's "temporarily permanent" bookmarks functioned every bit as well as he'd hoped they would.

Fry believed he had the makings of a great new product. Unfortunately, 3M execs were skeptical. In order to convince them, he provided a group of secretaries with blocks of the new notes and let them do with them as they pleased. Secretaries came up with more uses for these little yellow notes than anyone dreamed possible.

Fry then cut off the supply of Post-it® Notes. When the secretaries clamored that they wanted more, Fry had his proof that 3M had a winning product. The 3M execs were convinced and by 1990, Post-it® Notes were one of the five top-selling office supply products in America.

Notice how this story is told sequentially. The key points all take place as a sequence of events.

Category

If your story easily breaks into three or four logical parts, organizing it by category may be the best way to go. For instance, in this section of this book I am using categories to explain the different ways to organize a story. Categories will show your audience that you have thoroughly thought through all aspects of your content.

Bold Comparison

Any time you tell a story, one of your goals should be to make it memorable. By increasing the drama of your story, you will increase its memorability. One good way to increase the drama is to set up a bold comparison. You might compare the old way versus the new way... the wrong way vs. the right way... the slow way vs. the fast way.

For example, in his wonderful book, *Men are From Mars, Women are From Venus*, author John Gray creates a powerful bold comparison to explain how differently men and women think. As Gray points out, "Men mistakenly offer solutions and invalidate feelings, while women offer unsolicited advice and direction... Men tend to pull away and silently think about what's bothering them while women feel and instinctively need to talk about what's bothering them... Men are motivated when they feel needed while women are motivated when they feel cherished...." It may have been easier to write a book about the way men think or the way women think, but by comparing men to women Gray increases the drama and makes a difficult topic much more powerful and much easier to understand.

Another good example of Bold Comparison is the presentation I delivered when I worked for Broadway & Seymour, a software developer.

I started by explaining how banking systems used to be designed (the old way).

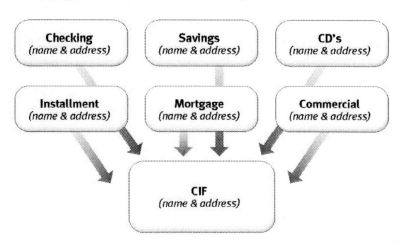

I showed how each account – checking, savings, and certificate of deposit – was separate, with name and address fields for each. Later, someone came up with the idea of a Central Information file. The file allowed bankers to see all the relationships a given customer had, but it didn't solve a basic problem. If a customer moved, the address needed to be updated in each account, one by one. This led to mistakes and inconsistency and was a nightmare for bankers.

At Broadway & Seymour, we did things differently (the new way).

Banking Systems - The New Way

We designed our system from scratch around the Central Information file, with address fields one time, one place only. Now, for the first time, bankers could see all the relationships that customers had, without the confusion of maintaining a separate CIF. Our system reduced errors.

For Broadway & Seymour, a bold comparison dramatically illustrated the advantages of the product. It increased the drama of our presentation and highlighted the value of our centralized CIF design.

Problem/Solution

One of the most often used and most powerful story telling techniques is the problem/solution format. In this format the storyteller starts by saying, "I had this terrible problem which was "X". I worked hard to solve it and created a wonderful solution which is "Y". Finish the story by talking about the results achieved. For example, if the Post-it® Note Story were told using problem/solution it would sound like this:

Back in the early 70's, a 3M employee named Art Fry had a problem. He was having trouble keeping bookmarks in place while singing in his church choir. During church services, Fry's bookmarks kept slipping out of his hymnal and falling on the floor. He knew there had to be a better way. Fry had an idea. He knew a 3M scientist had created a weak adhesive that would stick to paper, but not tear it. Fry took some of the adhesive and applied it along the edge of his bookmark. His solution worked perfectly and his church hymnal problem was solved! Fry now had a bookmark that could be temporarily glued into place without falling out. When it was time to move the bookmark, he could peel it off without damaging the pages of his hymnal.

Fry's bookmark worked so well he believed he had the makings of a great new product. But, when he tried to convince 3M execs about the merit of his peel-off note, they were skeptical. In order to convince them, he provided a group of secretaries with blocks of the new notes and let them do with them as they pleased. Secretaries came up with more uses for these little yellow notes than anyone dreamed possible. Fry then cut off the supply of Post-it® Notes. When the secretaries clamored that they wanted more, Fry had his proof that 3M had a winning product. The 3M execs were convinced and by 1990, Post-it® Notes were one of the five top-selling office supply products in America.

Step 2 - Visualize

So, now you have four techniques to organize the body of your story. But before it's ready, your story needs two other key parts, the tease and the transferrable lesson.

The Tease

One of the things I admire about the TV industry is its ability to quickly hook viewers. TV is a 50-year-old medium that has been refined over the years. The product TV offers is not its programming, but the audiences it delivers to advertisers. Because of this, TV producers have become acutely aware of what works to attract and retain these audiences. There's no better example of the right way to attract and hold viewers.

Shows like Prime Time Live and Entertainment Tonight always begin by teasing their top three stories: "Coming up, see exclusive footage of...", "Then we'll tell you three unknown facts about...", "And finally, we'll uncover secrets that help solve the mystery behind...". Rather than staring at a boring image of a talking head, the viewer sees footage from the upcoming stories while the host voices over the essence of the story and why you will want to watch. Why tease stories in this manner? Because it works! The TV producer is banking on the fact that if you are interested in one or more of the stories they tease, you will stick around to watch it.

What the TV world has learned is that you should hook your viewer in the first 10 to 15 seconds. This is foreign to most corporate presenters who make their audience wait several minutes before they get into any valuable content, and this after they have made the audience suffer through a long introduction. So the tease should be short and sweet. It entices me to stick around and answers the question, "why should I watch?" Let's look at how we might tease the main point of the stories we used in our previous examples.

Post-it® Notes:

How did a simple bookmark become one of America's top five office products? Stick around and you'll learn how a 3M employee turned necessity into a multi-million dollar innovation.

Categories:

Are you having trouble organizing your presentation? It may be easier than you think.... Coming up you'll learn the four main ways that work every time.

Men are from Mars, Women are from Venus:

Most men and women would like deeper and more satisfying relationships with each other, but they have no idea how to obtain it. In his book, *Men are from Mars, Women are from Venus*, author John Gray gives us remarkable insight into this challenging problem.

The Transferrable Lesson

The final part of your story is the transferrable lesson. It works to entertain people with stories, but chances are you want more than that. Now that you've hooked your audience and held them captive with a well-told story, it's time to come full circle. Make sure that you show your viewers how to apply this lesson to their situations. One good way to organize and communicate the transferrable lesson is with a list.

Lists

Here's an example of the power of lists. Several years ago I was in the Baltimore airport and was getting ready to fly back to Charlotte. I decided to buy a magazine to read on the way home, and walked over to a newsstand. There were over a hundred magazines to choose from, but because I was training for a marathon, the first magazine I looked at was *Runners World*. Before I picked it up, I noticed that in bold type across the top of the magazine it said, "Eight Ways to Increase Speed." I thought, "that's pretty exciting - I would like to go faster." I picked up the magazine and noticed that the bottom said, "32 ways to make any run easier." Sold! I didn't even open the magazine. Somehow the people at *Runners World* knew my exact pain. They knew I wanted to go faster and I wanted to make my runs easier. Then they communicated my pain in a catchy list: "Eight ways to increase speed."

Lists let the audience know, "Pay attention… you're about to get some good stuff." Lists also cause your audience to take notes. As soon as you say, "there are six key things you should know about…" the audience knows they are about to learn more things than they can remember, so they begin to write down the key points. When your audience begins taking notes, you know you are delivering value and you know they are actively listening.

Gaining Executive Buy In

Your education-based marketing program will be much more successful if you have executive management on your side. Executive buy in will usually bring you a bigger budget as well as more internal resources. There are three key tools that you can use to gain executive buy in; the concept paper, the business case and the use case scenario

Concept Paper

As its name implies, the concept paper is a short, 1-3 page description of your concept. It should include the following:

- Purpose of the program - a brief description of why are we doing this.
- Audience & challenges - who do we hope to reach and what are their biggest challenges

- Title – proposed name of the program. Ideally this should be short and catchy, but accurately convey why the viewer should watch.

- Segments – a description of the main topics that will be covered within the program. You may also want to include some sub bullets that indicate main points within each segment

- Call to action – what do you want the viewer to do after watching the program?

Business Case

Marketing organizations are being increasingly required to build a solid business case for their marketing objectives. A business case should not be a "pie-in-the-sky" scenario, but a genuine and realistic attempt to document the economic impact of the proposed program. A good business case should include seven key factors:

- Total Exposures – How many people do you intend to make aware that this program exists?

- Total Viewers – How many people watched the program?

- Next Steppers – How many people selected one of the next steps?

- Next Step Impact – What is the economic benefit to your company if some one takes the "next step"?

- Investment – what is the estimated total cost of ownership to develop the program?

- ROI – Based on the total cost of ownership, how much return on investment do you estimate this program will bring?

- Timing – How long will it take to develop the program and what period of time will be needed to get a suitable ROI?

<u>Use Case Scenario</u>

The final tool that you may want to use to gain executive buy in is a use case scenario. A use case scenario is a fancy way of saying, "how would we use this content if we had it?". For example, if you own a residential real estate agency, you may want to create EBM content that offers helpful advice on how to sell your home for the most money in the least amount of time. There are a wide variety of ways this program could be used. A use case scenario will examine these ways and allow management to better understand what you are proposing. In this case, the main way to use the content may be to put it on CD-ROM so that your real estate agents can give it to prospective customers. Agents are often interviewed by potential sellers before they select the agent they want to list their home with. Imagine how powerful it would be if, at the end of the interview, the agent turned to the home owner and said, "I hope the ideas I shared with you today are helpful… let me also give you this CD, it has everything we just discussed in a helpful TV show format, keep it as my gift." This gesture would obviously position the agent as a good choice, if not the best choice.

Also, real estate agents will tell you that their main source of new customers is referrals. If several months from now, a friend asked the home owner who they should list their home with, the satisfied home owner could pull out the CD and say, "Oh, you should use Sue, she gave me this great CD, You're welcome to have it, it's got her phone number on it if you want to give her a call."

Chapter Summary

Bottom Line

Step 2: Visualize the concept and the approach.

Develop a clear outline and approach for your content before investing in designing and building it.

How?

- Develop your concept
- Organize your content
- Create the tease
- Finish with transferrable lessons
- Use 3 tools to gain executive buy in – concept paper, business case & use case scenario

Thought Starters

1. Take a look at your current marketing initiatives. Can you take advantage of EBM and restructure your marketing program to build trust with your customer?

2. Can you list the exact pain or problem that you solve for your customers?

3. Do you have any projects that you can apply a business case and use case scenarios to gain executive approval?

CHAPTER 6

Step 3 – Design

- CONTENT BOOT CAMP
- THE 6 IMMUTABLE LAWS OF EDUCATION-BASED MARKETING
- ATTRACTING AND RETAINING TECHNIQUES
- SEGMENT MAP AND SAMPLE SCRIPTS

The MindBlazer 5-Step Creative Process

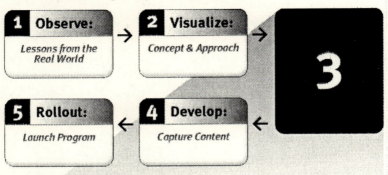

3 Design:
Content Design Bootcamp

1. Knowledge Transfer

2. Competency Capture
- *Story Packages*
- *Heroes, Luminaries & Experts*
- *Problem, Solution & Results*
- *Transferrable Lessons*

3. Attract & Retain
- *Titles, Teases & Hooks*
- *Lists*
- *Analogies & Metaphors*

4. Segment Map

5. Script Refinement

Step 3 - Design

> *"The world is so fast that there are days when the person who says it can't be done is interrupted by the person who is doing it."*
>
> *-Anonymous*

> *"Honest disagreement is often a good sign of progress."*
>
> *-Mahatma Gandhi*

> *"The journey is the reward."*
>
> *-Taoist Saying*

Content Boot Camp

In order to quickly and effectively capture corporate knowledge, we have found that a one-day, hands-on content boot camp provides a great way to jump-start the process. In the previous step, you create the organizational outline for the story, including tease, body and transferrable lessons. The goal of the content boot camp is to collect all the information (and knowledge) to complete the story.

The best size group for the content boot camp is about 5-10 employees with a variety of backgrounds. There is no magic number, but we find that you want at least a few people to attend and if you have more than a dozen it can get a little unwieldy. Your group should include the following types of people:

- Subject matter expert (for each topic to be covered)
- Marketing
- Sales
- Customer service
- Technology.

In addition to the client participants, MindBlazer brings the following three people:

- Boot camp facilitator
- Script writer
- Project manager.

No matter whether your "client" is an outside company or colleagues in your department, if you are organizing a boot camp, you'll want to include all these types of people.

Our boot camps usually start at 8:30am and wrap up about 4-5pm. We start the day with introductions. Specifically we like to know each person's name, what they do and how long they have been doing it. Once the introductions are done, I kick off the boot camp by sharing a presentation called the "Six Immutable Laws of Education-Based Marketing."

The 6 Immutable Laws of Education-Based Marketing

We ask participants to keep these laws in the back of their minds as we capture stories and develop the content. The "Immutable Laws" presentation serves two purposes. It's a great way to quickly bring our client up to speed and get the boot camp team excited about the potential of what we are about to do. It's also a powerful way for us to build credibility. The more credibility you have, the more time the client will invest in educating you about their content. If you are organizing your own boot camp, you'll want to get participants invested in the process early.

Our laws are presented in a colorful PowerPoint presentation and cover the following points:

The First Law, The Law of Viewer Value

> *If you want to create meaningful content that is valuable to the viewer, focus single-mindedly on solving the viewer's biggest challenge. In other words, "talk about what your viewer wants to learn, not about what you want to sell".*

The Second Law, The Law of The Story

> ***1. Stories are a powerful audience retention technique***

2. Stories are interesting, easy to understand and easy to repeat

3. A good story has four key parts:

    Problem

    Solution

    Results

    Transferrable Lesson

The Third Law, The Law of Three Perspectives

To add credibility, let the audience learn about your topic through three perspectives:

    Hero

    Luminary

    Expert

(We'll discuss this idea a little later in the book.)

The Fourth Law, The Law of Hooks, Teases and Titles

1. If your viewer doesn't experience value in the first 15 seconds they will click away ... to hook them, use meaningful titles and tease your top stories.

> **2. Good titles and teases answer the question, "why should I watch?"**

The Fifth Law, The Law of Metaphors

> **We become familiar with the unknown by relating it to things we do know ... metaphors help us create a mental map.**

(You'll learn more about this in a few pages.)

The Sixth Law, The Law of Lists

> **Viewers place high value on lists. To add value to your content, consider organizing the transferrable lessons into a list.**

Once we have educated the audience and built credibility, it's time to begin identifying our client's competency and building story packages. Remember, story packages are short pieces that cover a particular topic. If you have watched a news magazine TV show you have seen a story package. The host of the show usually says something like, "our first story takes us to…". The host then tosses to a reporter who has developed the package. The reporter's job is to make the story compelling, easy to understand and applicable to the viewer.

Unlike a typical corporate presentation where a subject matter expert delivers all the content, a story package usually involves interviews with three types of people:

- Heroes
- Luminaries
- Experts.

Heroes are people who have done what you are trying to teach. This is typically a customer who had a specific challenge, worked diligently to implement a solution and experienced positive results. When we mention heroes, clients often say, "Oh, we use testimonials all the time."

Please note, there is a big difference between a testimonial and a hero story. In a testimonial the customer talks about how wonderful your company is. While these are nice for patting you on the back, they usually don't add much. Anyone will say nice things when you ask him or her to and even worse, you don't learn anything from a testimonial.

In a good hero story, you should learn what kind of problem the hero had, then you learn how he or she went about implementing the solution. Ideally the hero will share "pearls of wisdom" gained through experience. These "pearls of wisdom" are part of the knowledge you are trying to capture and share.

Finally, the hero should relate the specific results achieved and the results you can expect to achieve if you implement a similar solution. Ideally we like to build the story package around the hero. We find that heroes talk in terms that other prospective customers can understand. Also hero stories inherently increase the drama of your story. Viewers get pulled in and become curious to see how the hero dealt with the same challenges that they face.

Next the story package should include comments from a *luminary*. Luminaries are third-party experts who don't work at your company. Theoretically they are unbiased and generally they are able to give "the big picture" as well as good solid advice. Luminaries may have written books about your topic, been featured in magazines, or may work for trusted research firms.

Finally the story package should include comments from your company's *subject matter expert* or *experts*. We find the best role for the expert is to punctuate what the viewer learned in the hero story and offer helpful advice about the category.

Like a good newsmagazine TV show, a host serves as the "glue" to hold the story packages together. The host provides the teases, the story set up and any concluding or summary comments.

Here's a tip when you are selecting people to include in video interviews. Look for people with energy. You want people who are passionate about their topic and are willing to talk about it with zeal. Unfortunately, one of the drawbacks of video is that it tends to diminish people. Some experts estimate that video diminishes people by up to 30%. This means that if someone delivers a high-energy presentation during the interview, when you watch the videotape, it will appear as normal energy. If someone delivers a normal-energy presentation, it will appear a bit dull. If someone delivers a dull presentation, you're dead.

During boot camp, we build the content by first identifying and learning about good hero stories. Understanding and capturing full story packages usually takes the remainder of the morning. Throughout the day, we project our notes onto a screen so that everyone can see what we are capturing and minor corrections can be made.

Attracting and retaining techniques

In the afternoon we shift gears and work on attracting and retaining techniques. Now that we know the story packages that will be included in the piece, we can develop effective titles and teases. Developing titles is a simple exercise of brainstorming a dozen possible titles and then letting group members vote on the one they like the best. The same is true for teases. We like to suggest possible teases and lists in our initial content outline. During the boot camp, we like to further develop the teases and the lists. The content outline is the skeleton. The boot camp lets us put some meat on the bones.

Finally we want to go back to the stories and brainstorm a list of lessons that can be learned from each story. These lists become powerful sets of transferrable lessons. Nothing is more powerful for positioning you as a trusted advisor than wrapping up a story by having your expert say, "As you can see from this story, there are seven key things that you should know. Let me share these with you."

Segment map and sample scripts

We like to wrap up the content boot camp by presenting a segment map and a sample script. We recommend breaking a client's content into logical segments. No segment should be longer than five minutes. A segment map is a high-level view of what will be covered in each segment, including topic, title, tease, hero story, luminary, expert and transferrable lesson. To further underscore what the content will look like, we usually have our writer read a rough draft of a sample segment. There might be three to five story packages in a segment, although sometimes one or two is all you need.

If done properly, the client will be amazed at how much content was captured during the day. We are usually asked for a copy of the notes. Our clients are also impressed that we are able to take the content and shape it into a coherent message so quickly.

If you're running a boot camp on your own, we advise choosing software with a good outlining tool. Project your outline on a screen. You'll start with preliminary content - the bare bones of your subject. For instance, when working with a major pet food manufacturer, we developed the topic "Six Signs of Optimal Pet Health" To develop the outline during boot camp, we simply said, "Tell us about the six signs. What are they? How can pet owners spot them?" We heard about such concepts as good body condition versus bad in a pet. We wrote down every applicable idea so it became part of the outline.

Attendees saw that outline take shape. We built a document with all of their thoughts. What began as a half-page document soon grew to ten pages. It was powerful for attendees to see their thoughts on the screen, and understand those ideas would become part of the finished script. At a successful boot camp, attendees will leave excited about the project.

The last thing that needs to be done in this step is to complete and refine the script. The script will become your blueprint.

One word about language in the script. Your viewers won't watch if you sound too commercial. The pet food company's script would say, "When you're shopping for a nutritional pet food..." not, "When you buy Brand X...." Don't name your product over and over. Simply by referring to your category, rather than your particular product, you will subtly position yourself as the leader.

Chapter Summary

Bottom Line

Step 3: Design your content

Employ a consistent design process that will give you a compelling show each and every time.

How?

Make sure that you bring the right group of people together for a content design and brainstorming session. Use this session to set the stage for how the team will design content. Make sure that you collect good stories and data on heroes, third-party luminaries, and experts on your service or product. Ensure that your sponsors review your final script before you fire up the production engine to build the content.

Thought Starters

1. Who are the unsung experts in your organization? How would you package their knowledge into stories for your customers' benefit?

2. Who are the key stakeholders (product experts, customer service teams, etc.) in your company that could contribute to a content design session focused on earning your customer's trust?

CHAPTER 7

Step 4 - Develop

- **CAPTURE CONTENT**
- **PRODUCTION AND EDITING**

The MindBlazer 5-Step Creative Process

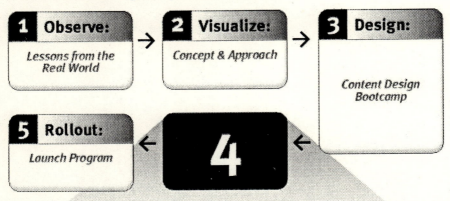

4 Develop:
Capture Content

1. Produce
- *Interview Heroes, Luminaries & Experts*
- *Shoot Talent*

2. Post Produce
- *Develop Professional Graphics*
- *Edit Video • Client Review & Refinement*

3. Adapt for Distribution Channels
- *Internet • CD-ROM • Print • Verbal*

Step 4 – Develop

"There are no shortcuts to any place worth going."
-Anonymous

"One that would have the fruit must climb the tree."
-Thomas Fuller

"Even if you are on the right track, you'll get run over if you just sit there."
-Will Rogers

Capture Content

With script in hand, it's time to begin capturing content. This step includes producing, editing and adapting your content for a variety of channels.

Let me explain what I mean by adapting your content. If done properly, what you will be creating is a "digital asset" that can be distributed through a variety of channels. The three most logical channels for distribution are:

- Video on-demand from your website
- CD-ROM at point of sale or direct mail
- Verbally, by sales rep.

Your content for the first two channels, web and CD-ROM, will be identical. The only difference is the delivery vehicle, streaming over the Internet vs. stored on a CD-ROM. The content for verbal channel, as I like to call it, is a bit different. The message is similar, but since it has to be delivered verbally without the aid of video, the content is a bit simpler. In adapting for the verbal channel, the two most powerful script components are stories that can be retold and lists.

A great example of how stories can be adapted for all three channels is the content we developed for Georgia-Pacific. The Consumer Packaging Division of Georgia-Pacific was in the process of completely revamping its Technology and Design Center. In addition to renaming it The Innovation Institute, Georgia-Pacific also overhauled the center's structure and developed clear compelling graphics to tell their story. But they wanted more. Georgia-Pacific wanted to use the power of story telling to capture past experiences of how they helped consumer-packing clients. Georgia-Pacific also wanted to turn those experiences into a set of consumer packaging laws they could share with customers and prospects.

Our solution was three-fold. First, since the dominant way customers would experience this content was face to face, we focused on creating a series of PowerPoint presentations with embedded video story packages. We were dealing with some complex topics like "Supply Chain Collaboration," "Packaging System Optimization," and "Shelf Velocity." In order to make these easy to understand and easy to remember, we created a metaphor for each.

We all become familiar with the unknown by relating it to things we know. Metaphors help us create a mental map. For example, when the car was first invented, it was called a "horseless carriage." Before there were cars, horses pulled carriages. Instead of explaining that the internal combustion engine would negate the need for a horse, it was easier to explain the car was a carriage that could go without a horse. The horseless carriage metaphor was easy to understand because the description was made up of two things everyone was familiar with.

So a metaphor helps you put things where they belong. Before we can understand something, we all have to put things somewhere in our brain. If you have a topic that is complex or difficult to understand, I recommend creating a metaphor.

In addition to using the content for live presentations, we also created a self-contained CD-ROM version of each presentation that Georgia-Pacific gives to visitors as a way to take some of the lessons they learned at the Innovation Institute home with them. All of the content covered by the live presenter at The Innovation Institute is delivered by a professional host on the CD-ROM version.

Georgia-Pacific made the content available not only on CD-ROM but also on-demand from their website. In some situations, a customer might have an immediate challenge and not have time to set up a visit to The Innovation Institute. The web version allows a Georgia-Pacific engineer to get on the phone and direct the customer to the website. During the conversation, the Georgia-Pacific engineer can say, "To teach you some of the principles of how to improve shelf velocity, I'd like you to watch a short video... just click this link...". By making content available via streaming video on the Internet, Georgia-Pacific has extended the walls of The Innovation Institute to anywhere their customers are.

Production and editing

Now that you understand the power of creating content that can be distributed in a variety of channels, let's go back to where we started this chapter and talk about production and editing.

During the content boot camp, one of the tasks was to identify the best heroes, luminaries and experts. Once your script is developed and approved, it's time to contact the various heroes, luminaries and experts and arrange for video interviews. If they understand you are not looking for an endorsement, but merely some helpful friendly advice on a topic that they are familiar with, most people are more than willing to help.

I recommend hiring an experienced TV and corporate video shooter. You should be able to shoot about three good interviews in a day. There are two important things you should do to capture good interviews. First, make sure the interviewee understands what you are trying to achieve and what kind of content you are trying to create. To help interviewees out, we often write sample quotes into the script. The second thing you can do to ensure a good interview is to be a good audience. Just like any conversation or presentation, you tend to draw a lot of your energy from your audience.

Don't make your interviewee look into the camera. Have the video shooter position the camera over your right shoulder. To help your interviewee relax, encourage him or her to look at you, talk to you, and pretend the camera isn't there. To get things started, you might ask some "dummy" questions -- questions you don't really care about, but will get the subject talking. Then ask your real questions, encouraging the subject to share helpful advice. If your interviewee is still nervous, don't worry. After a few minutes of questions, we find that most people will relax and move into a comfortable rhythm.

Prior to the interview, write out the questions you will ask. For instance, in our Lowe's interview, we asked the experts to finish these sentences:

- "The most important thing to remember when you renovate a kitchen is...."

- "The biggest mistakes people make when they renovate their kitchens are..."

- "The best thing you can do to dream about your ideal kitchen is to..."

By asking the interviewee to finish the sentence, you tend to capture more complete thoughts. If your interviewee talks in fragments and merely answers your questions without completing the sentence, you can shoot a host who can ask the question, then edit footage of your interviewees responding.

When you're done with the taped interviews, you will have an hour or two of videotape. Pick the best sound bites and then shoot the host. The host

Step 4 - Develop

will provide the teases, intros and summaries. The host can also provide the segues that glue the interviews together to form a tight package.

When editing, I recommend that you keep your segments to five minutes or less. My experience is that you should not ask anyone to watch something on the Internet or CD-ROM for more than five minutes. If your content is longer than five minutes, break it into a set of logical segments. For instance, a program on renovating a kitchen might be broken into three five-minute segments: one on planning, one on design and one on installation.

Chapter Summary

Bottom Line

Step 4: Capture Content

Capture your content and stories using the power of metaphors. Make sure that you prepare your content for distribution in a variety of channels: Video on- demand over the Internet, CD or DVDs, in store verbal (where applicable), and print.

How?

Take the time here to capture content that can be deployed in multiple channels. This step includes production, editing, and adapting the content for multiple delivery channels.

Thought Starters

1. After you design your content, ask yourself the following questions from the point of view of your customer:

    a. Why would I want to watch or read this?

    b. What have I learned? And what can I do with this new found knowledge after I watched this?

CHAPTER **8**

Step 5 - Rollout

- **LAUNCH PROGRAM WITH SOUND TESTING AND TRAINING**
- **KEYS TO A SUCCESSFUL ROLLOUT**
- **QA AND REVIEW**
- **MULTI-CHANNEL LAUNCH – COORDINATED ON AND OFFLINE**
- **LESSONS FROM WACHOVIA**
- **TRAIN YOUR STAFF**
- **MONITOR AND REFINE YOUR PROGRAM**

The MindBlazer 5-Step Creative Process

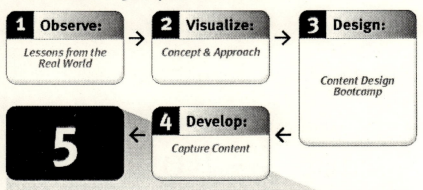

5 Rollout:
Launch Program

1. Test Launch
2. QA & Review
3. Training
4. Coordinated Launch Across Multiple On-Line & Off-Line Channels
5. Monitor & Refine Program

Step 5 – Rollout

"Come to the edge He said.
They said: We are afraid.
Come to the edge He said. They came.
He pushed them, and they flew..."
-Guillaume Apollinaire

"Only the spoon knows what is stirring in the pot."
-Sicilian Proverb

"God is in the details."

-Mies van der Rohe
"New York Times"
August 19, 1969

<u>Launch Program with Sound Testing and Training</u>

Step Five is easy to overlook. But we have found that the testing, QA and training phase is one of the most critical aspects of a successful program. We run parallel paths. As the content is being developed, we also begin testing and refining the technology part of the education-based marketing program.

Once you have everything built and ready to go, it is time to roll out your program. Many people make the big mistake of not paying attention to how the audience is receiving the program. Why does this happen? There are three main reasons. First, often times most of the project time is spent building the content, leaving very little time to test the program and content before rolling it out to customers and prospects. Second, excitement to launch causes many people to rush their programs because they just can't wait to see this new marketing tool on their websites and CD-ROMs. Third, they didn't have a method that gave them a systematic way to roll out their programs. Do not make this mistake. A poor rollout will negate all the hard work that you have put into building a beautiful program.

Keys to a successful rollout

Use these five steps to roll out your program successfully:

- Test
- Review
- Train staff
- Multi-channel launch
- Monitor and refine your program.

Let's look at a case study of a successful rollout. MindBlazer and Wachovia Bank partnered to build Money Minutes TV. The goal was to provide a portion of Wachovia's customers with helpful educational programs to improve their wealth management, retirement planning, and estate planning. Their research showed that their customers' biggest challenges were threefold:

- Simplifying daily money management
- Best ways to fund large purchases
- Adequately funding retirement.

These challenges were ideal for creating compelling content. Our recommendation was to create one segment for each challenge. Wachovia also told us they were interested in using this content to both deepen existing customer relationships as well as attract new customers to the bank.

Prior to starting the project we developed some rough marketing ideas. To deepen customer relationships we recommended the following strategies:

- Create a CD-ROM version to distribute at branches, in statements, and attached to ads
- Put "Money Minutes TV" icon on the home page
- Put "Money Minutes TV" icon on the online banking page
- E-Mail customers with a "click to learn more about...." message
- Promote "Money Minutes TV" at ATMs.

In order to position "Money Minutes TV" as a tool to attract new customers we recommended the following outreach programs:

Step 5 - Rollout

- Conduct a PR campaign

- Direct mail CD-ROMs to non-customers

- Send e-mail link to opt-in lists

- Advertise "Money Minutes TV" in *Money* magazine with CD-ROMs as an insert for the magazine

- Advertise "Money Minutes TV" on TV

- Insert CD-ROMs into *Money* magazine.

After discussing these ideas Wachovia agreed that there were a variety of good ways to leverage this content for both customer move-up strategies as well as attracting new customers.

QA and review

Now let's skip ahead to when the program was completed. Our program design team reviewed it with the following criteria:

Critical learning elements

Make a list of the critical learning elements that your content is designed to convey. Provide this to your internal quality group and ask them to check off each critical learning element as they view the content.

PSRTL - Problem, Solution, Result, Transferrable Lesson.

Every story you tell should follow the problem, solution, result, and transferrable lesson formula discussed in this book. Watch your content to ensure it does indeed follow this formula. It is very easy to leave out the transferrable lesson part of your story, so watch specifically for that.

Look and feel branding

Review the program for subtle cues it will deliver. Does it support the branding goals that you are trying to achieve? Will its intended audience relate well to the people used to tell the stories? You may find that one luminary you selected, for example, may not be as dynamic on screen as you expected. Maybe your test audience just doesn't care about your hero. Or perhaps the people in the program aren't diverse enough.

Production value

The show must be well produced in order to attract and retain your audience. Scenes must be well lit. A poorly lit video will not be watched. The audience will just tune out or click away.

The show must be shot effectively. A few types of shots are standard. You'll likely begin with a wide establishing shot to let the viewer know where the segment takes place. A medium shot will show who is there. Then you'll go to tight shots -- close-ups of one person at a time. Let your subject fill the screen. Remember, a computer screen is generally much smaller than a TV screen. Wide or even medium shots of people can be very difficult to see on a computer. Most of your content should be tightly shot.

Shoot people from the chest up. If you include the whole body, faces will be difficult to see. You also want to minimize movements in the background as well as fast panning shots. Fast-moving video will be more demanding on both bandwidth needs as well as computer processor speed, and will actually provide a poor viewing experience over the Internet. Overall, you want your scenes captured "tight and bright."

Clear call to action

What is the purpose of the program? Does there need to be a call to action? Any call to action should have a clear purpose and be valuable to the audience. For example, to ask the audience to register and provide contact information without a compelling offer would detract from your program. However, allowing viewers to contact you for helpful follow-up information on their own schedule could add to the program.

Though you may resist doing more editing at this point, we urge you to refine the program if needed. Though a new shoot might cost another $2,000, that fee is much less than the cost of an ineffective program.

Review with key sponsor

After the internal test, it's time to unveil the program to your key customer or sponsor. This is very much like delivering a custom-tailored suit. Just before the tailor delivers your custom suit, you will be asked to come in for a fitting. The suit is pretty much done, and the tailor wants to review it with you and test the fit before he makes the final touches and delivers the custom suit. Likewise, when Wachovia reviewed their program with us, they came up with a number of enhancements and change requests. We were able to work as a team to further improve the program. The completed program then survived an extensive review by various members of the Wachovia project team.

The initial review by the key sponsor saved us a tremendous amount of time. Just like the tailor, we were able to do a fitting with Wachovia before finishing up the program. The final program was then easily approved by the rest of the team after the key sponsor's insight was incorporated.

Multi-channel launch - coordinated on- and offline

Now that your content is ready to be launched, consider the trend towards coordinating your online and offline marketing. We call this integrated or multi-channel marketing. Your program's value is greatly enhanced if your content is delivered via multiple channels. The right multi-channel approach will allow you to provide a consistent marketing message to all your customers. You will be able to touch a single customer multiple times with the same message over different mediums, resulting in a larger impact on your customers.

Wachovia Bank maximized their program by launching it using multiple channels, online as well as offline. Wachovia will be delivering the content to customers via CD-ROMs at bank branches in Florida. The bank will also direct mail CD-ROMs to a large number of their customers. By using existing communication channels to market and deliver this content to their customers, the bank will reap maximum results.

Lessons from Wachovia:

Key lessons that you can learn from Wachovia are:

- Coordinate your message and launch.

- As you make the program available over the Internet, offer the same content through traditional marketing channels.

- You could direct mail CD-ROMs to customers as well as invite them to view the online program via email and other online traffic generating methods.

- If you are a retail operation, make these available as gifts to your customers at your retail locations.

- Leverage your media buys that are already in place. For example, if your marketing plan includes a full-page advertisement in *Money* magazine, consider providing a CD-ROM insert and attach your program to the full-page advertisement that you have already budgeted for.

Train your staff

Wachovia is also training its staff to deliver a short verbal seminar, giving customers value and insight into the content before offering the CD-ROM to customers. With that training, employees will be able to successfully introduce Money Minutes TV to their customers at branch locations. Many companies with excellent marketing programs do not achieve their goals because their staff lack the knowledge of how to use the tools provided. Wachovia Bank maximized the value of its program - and supported its staff - by implementing the following steps:

- Adopted a multi-channel approach to make sure that customers had more than one way to view the program.

- Made its call centers ready to provide assistance to customers who might have trouble accessing the program using the Internet.

- Provided CD-ROMs as a back-up. If a customer called the toll-free number for help, the customer representative could send a CD-ROM to the customer as opposed to trouble shooting Internet problems over the phone. This was designed to ensure that customers get as easy an experience as possible when they view the content.

- Trained employees on the key learning points of the program so they would understand how the program could help customers.

Monitor and refine your program

Closing the loop on your marketing program is vitally important. Yet most companies do not allocate the necessary resources and time to come back to review and fine-tune programs to achieve their stated objectives. In his book *The 7 Habits of Highly Effective People,* Dr. Stephen Covey talks about how important it is to sharpen your saw for better results. Closing the loop is very much like sharpening your saw. You need to press the pause button, observe whatever results you are achieving, fine-tune the program, and then press the play button again to continue with your campaigns.

Our pet food client knew that only 20% of their customers were visiting their web site, and therefore 80% of their customers (or 4 out of every 5) were not able to access the content. After the online video worked so well, they closed the loop by exploring offline distribution channels to produce even better marketing results.

Chapter Summary

Bottom Line

Step 5: Rollout and Launch

Make sure that you have done adequate testing of your technology and infrastructure, as well as verified all aspects of your content for accuracy and playback.

How?

This chapter covered the 5 steps necessary to roll out your program successfully: Test, review, train staff, multi-channel launch, and close the loop by monitoring and refining your program.

Thought Starters

1. Can you articulate the specific goals and metrics of your current or upcoming marketing campaign? How will you assess your results? How will you communicate these results to your various company stakeholders?

2. Is your company's field sales staff trained to handle questions that might be generated due to a successful marketing campaign?

Part IV

The Technology

CHAPTER 9

Technology of Story Telling

- **WHY TECHNOLOGY IS IMPORTANT**
- **DELIVERY PLATFORMS**
- **ONLINE DELIVERY**
- **CLOSING THE LOOP – TRACK AND REPORT**

Closed-Loop Marketing Platform

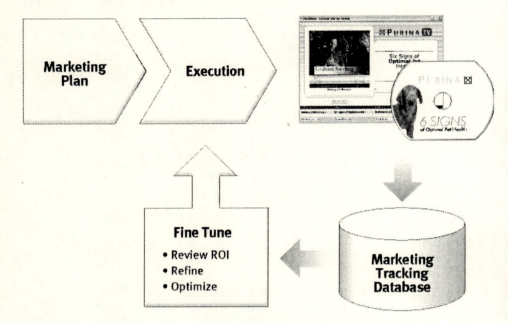

Technology of Story Telling

"Any sufficiently advanced technology is indistinguishable from magic."

— *Arthur C. Clarke*

"If it keeps up, man will atrophy all his limbs but the push-button finger."

-Frank Lloyd Wright

"The trouble with our times is that the future is not what it used to be."

-Paul Valery

Why technology is important

Technology can add significant value to your overall strategy. In the past, the tools for education-based marketing were typically seminars, print brochures, print newsletters, and annual reports sent through the mail. Now, you can increase your reach significantly with the use of technology.

You'll see gains in four ways. You will eliminate the barriers of time and space, expand your reach, create a scalable solution, and achieve consistency in your communications. All these changes can help you educate more prospects and increase your profits.

Eliminate the barriers of time and space

When you delivered a traditional marketing seminar, it was typically at a set time, say 10am, and at a set location, say the Marriott Hotel downtown. If your important customer, John Smith, was unable to attend at that time, he would have to wait until the next time you held the seminar. Alternatively, you could pay him a visit and conduct a one-on-one seminar. Once the seminar was over, the content was also over. You could not access that content again, unless you performed the seminar again. Now, because of technology, you have the ability to break down these barriers of time and space. Here's how:

Everything we've discussed so far deals with creating content that can be delivered to a remote audience. Using the appropriate software, digitize your program and make it available over the Internet. (We'll show you how later in this chapter.) Now if John Smith is unable to attend your seminar at the Marriott, you can still help him gain access to your seminar. You can email him the link to your online program, and invite him to come at his convenience. You have broken the barrier of time by allowing him to watch your seminar when he wishes. You have broken the barrier of space by allowing him to watch at his home or office, over his computer.

Increase your reach

You used to reach only the people who came to the physical seminar. Now you can extend your reach beyond your city, your state, even your country. You have now made your program available 24 hours a day, 7 days a week. We refer to this as 24 by 7 access. And because of this reach, you are now able to distribute your content and programs even while you sleep. Imagine if you have customers in Asia who wish to access your content. The sun rises in Asia as it sets in North America. Your Asian clients would be accessing your content as you are reading bed-time stories to your children.

Scale the solution as your business grows

You now have a scaleable solution. "Scaleable" means that you can grow your program based on your needs. For example, if you wanted to reach 1,000 people, you would have to get a bigger hotel room. If 60,000 people wanted to hear you, you would have to rent a stadium. And how would you accommodate a million people? This is the first time that you can provide access to a million people and not have to worry about getting a larger stadium. As more and more clients access your content, you grow the number of computers or servers in your network to provide access to more people.

Consistency throughout your communications program

Usually with greater reach, you lose consistency. As you use different people to distribute a message across multiple geographic areas, your speakers will all add their own personal nuances to the message. It was very difficult to ensure a consistent message using the old way as your messengers spread the word. Now you can record this content and make sure that your message is delivered consistently around the world, 24 by 7.

Delivery platforms

There are so many variations and combinations of technology that your choices are almost endless. I want to give you some tips on making the right choices without having to hire a gear head to figure it out for you. First let's consider your delivery platforms, online and offline. Online refers to using the Internet to make your content available to anyone who is allowed to view your website. Offline can include CD-ROMs, DVDs, and VHS tapes. Let's talk further about these offline platforms.

CD-ROM

Today, most computers and laptops ship with the minimum of a CD-ROM drive built in. Many even have DVD-CDROM combination drives built in. Software programs have become so large that most software providers ship their software on CD-ROMs, causing hardware manufacturers to include CD-ROM drives as a minimum configuration for their computers. CD-ROMs are very inexpensive to manufacture in bulk, light in weight, and can be packaged attractively. Research shows that 60% of American households have computers. This is expected to grow to 90% by 2008. Almost 100% of these computers are already equipped with CD-ROM drives.

DVD

DVDs have also become mainstream of late. I was pleasantly surprised that our research indicates that DVDs are being adopted at a very fast rate. This rate accelerates as more computers are bundled with DVD drives. Let me share some of the research with you:

- DVD Entertainment Group: DVD player sales in North America have grown from 9.8 million in 2000 to 16.7 million in 2001 to an estimated 20 million in 2002. The group estimates that nearly half of all US homes had DVD capability by the end of 2003.

- IRMA (International Recording Media Association): IRMA forecasts that by 2008, annual factory shipments of DVDs globally will approach 7 billion units.

- Adams Media Research: Projects DVD players in 91% of US Homes by 2008.

DVDs provide the highest quality of digital video and sound available to you today as a media for physical distribution of content. DVDs also provide up to 9GBs of data storage or about two hours of high quality video content.

VHS

Finally, we have VHS tapes. VHS is a medium that most American households have. It has been the main platform of choice for direct marketers trying to get video into the hands of their target market.

How you choose

So how do you choose from these offline delivery platforms? The table below provides some examples on various things to consider when choosing your media delivery platform.

Target Market	**Usage**	**CD**	**DVD**	**VHS**
Consumer	Office	Yes	Yes	
Consumer	Home		Yes	Yes
Employee training	Office	Yes	Yes	Yes
Employee training	Home		Yes	Yes

The main variables are:

- **Your target market**: Are you targeting people who will have the ability to watch your content on the medium that you have chosen?

- **Usage**: Consider what type of content you are trying to deliver. Is this something that is easier to watch at home on a VCR? Or is this something that is work-related and therefore easier to watch on a laptop computer?

Weigh all these factors together. Remember, it makes no sense to create a DVD, even if it looks and sounds beautiful, if your target market cannot play that medium.

Online delivery

Now that you are ready to deliver your content online, there are a few things to consider.

Delivery method

There are two ways to deliver your content online. You need to either stream the content or deliver the media file to the viewer's computer. The table below summarizes the pros and cons of both methods:

Delivery Considerations	Download	Streaming
Files delivered to viewer's PC	Yes	No
Viewer can edit video files	Yes	No
Need to remove media files	Yes	No
Retain control of video files	No	Yes
Buffering (waiting for video)	No	Yes
Need high speed internet	No	Yes
Must be online to watch	No	Yes

Delivering media files has the advantage of allowing viewers to see the content after it has been downloaded, even if the viewer is not connected to the Internet. However there are two major disadvantages:

- Viewers have to delete the files later. Some viewers won't like having files left on their computer.

- You relinquish control of the video file to the viewer. These two disadvantages make streaming the content the more popular choice.

Let's talk about streaming. To receive a stream, your viewer must be online, connected to the Internet using high speed or broadband access. To help us understand how streaming works, think of a water pipe connecting the city's water supply to your house. The city has large water pipes that deliver water to your district or community. Then there are smaller pipes that connect from your community's distribution point to your house. When you turn on your tap, water flows through these pipes into your home.

Likewise, the Internet is the pipe that video content is streamed through to get to your computer. When you click "play", content flows through these pipes just as water flows when you turn on your tap. Video content tends to require larger pipes or higher bandwidth. That's what we call broadband or high speed Internet access. You can access video content with a dial-up line at 56kps. However, the experience would be so poor it would discourage you from viewing the content. The images will be grainy and distorted, and the video might stop and start in a jerky fashion while it's running.

Broadband is a prerequisite to having a positive experience viewing video online. Broadband access includes services such as cable modems, digital subscriber lines (DSL), and T1 or T3 access.

Encoding

Video files take a lot of space. They are too large to send over the Internet and will clog the pipes. Encoding is the process of compressing video files into a significantly smaller format for the purpose of streaming over the Internet. Even though these encoded files are significantly smaller than the original digital video file, they are still rather large. There are a myriad of encoding choices. The higher the speed, the higher the video quality and the larger the bandwidth necessary for viewing the content. We recommend that you encode at both 100kps and 300kps speeds.

Media player platforms

Next, you need to consider which players you wish to support. There are currently three major player platforms: Windows Media Player (WMP), RealPlayer (RP), and QuickTime (QT). When you encode video, you are actually encoding it for a particular player. For example, if you wanted to make your content available with WMP, you could choose to encode it for playback at two speeds - 100kps and 300kps. You now have two files sitting on your video server. If you also choose to support RealPlayer at the same two speeds, you will need to encode the video again, at those two speeds, for the RealPlayer platform as well. In order to support three platforms at two streaming speeds, you will actually end up with six encoded media files of the same video on your video server. If your intended viewer does not have the right player already installed on his or her computer, any of the three player platforms can be easily downloaded at no charge from their suppliers.

As you think about which media players to support, note that WMP has been bundled into the Microsoft Windows operating system since Windows 2000. WMP has become a very safe platform to encode for since Microsoft has 95% market share of operating systems for all personal computers. WMP also works on the Apple Macintosh operating system. The current trend is to support all three player platforms for content intended for the consumer mass market. If you support all three, you'll help ensure your target audience can view your content.

There are other players available as well, some with higher quality playback and even better features. However, the downside would be that more of your viewers would have to download a special player. This download would be a barrier to watching your content, and many will click away without bothering to download the player. We recommend that you minimize the number of barriers to your content.

<u>Closing the loop - track and report</u>

Many websites are still asking visitors to register before allowing content to be watched. This is a very bad idea. Why? Registration represents yet one more hurdle in front of the potential viewer. Even if the visitor does register, people will often use false names and contact information to avoid adding to the barrage of email they are already receiving. It is much better to receive fewer contacts who are truly interested in your offering rather than a large number of contacts who aren't. Here's a good rule of thumb. Ask for contact information only after viewers have received enough value that they can imagine how speaking with you could be beneficial to them.

Allow your content to be viewed free of registration and other encumbrances. Then include a call to action, where viewers may register or ask to be contacted because of a specific offer they are interested in. This way, you will receive maximum viewership of your content and the registrations of only those people who are truly interested in your offerings.

Often, clients tell us they want to gather as many contacts as possible. We believe gathering thousands of unqualified contacts is unproductive. You'll devote far too much time culling through all those names to find the people who are genuinely interested in your product or service and able to buy. Having fewer and better quality contacts will significantly reduce your company's cost of sales and give you more time to spend on people who are truly intrigued by what you can provide.

Before you spend time collecting and reporting on data, make sure there is an action that can be taken regarding any metric that you track and report on. In other words, do not waste time tracking information for pretty reports that cannot help you make decisions and take actions to improve your programs. We call these Actionable Metrics.

There are six Actionable Metrics that I suggest you track and review:

1. **Marketing Response** - number of people made aware of your program. Start with recording the baseline of your marketing programs. How many people were made aware and invited to your content? This number grows on a monthly basis. It represents the total number of people that you invited, across all your campaigns and channels of communication.

2. **Conversion to Viewers** - how many clicked to view. How many visitors actually clicked to view your program after arriving at your site? The goal here is to understand if you did a good enough job teasing your content in order to attract the viewer.

3. **Content Selection and Usage** - which assets were viewed and for how long. How long are your video segments being viewed? If someone clicks away quickly, your content may have been titled incorrectly, or you may have to rework your tease. You can also review a summary to understand which video segments are being watched and which ones are being ignored. Take action to improve video segments that are not being watched.

4. **Call To Action** - How many people clicked on your call to action. Look at how often people are actually clicking on your call to action. Review this and improve your offer, placement, design, or clarity of the call to action.

5. **Clicked to Buy.** You may have a "click to buy" function. If you do, review the results. How many people actually showed an intention to purchase your offering by clicking on this button?

6. **Conversion to Customers** - How many visitors and viewers actually become customers? The goals are to analyze customer data to determine newly acquired customers as well as existing customers who purchased additional products and services.

Chapter Summary

Bottom Line

Technology can eliminate the barriers of time and space, allowing you to increase your reach geographically as well as stretching your reach to a 24 hours a day, 7 days a week program. You should also track your program's performance so that you can fine tune the program periodically.

How?

This chapter covered the various types of technology and methods to achieve a 24 by 7 program.

Thought Starters

1. How can you leverage the trend that more websites are migrating to being video centric versus text centric? How can your company take advantage of this trend?

2. How can technology eliminate barriers of time and space for you?

CHAPTER **10**

Lead Capture

- **THE LAST MILE OF MARKETING (OR FIRST MILE OF SALES)**
- **THE TRADITIONAL PROBLEM**
- **THE EBM SOLUTION: TURN QUESTIONS INTO SALES LEADS**
- **EBM LEAD CAPTURE BENEFITS**
- **EBM DIFFERENTIATOR**
- **LEAD CAPTURE FLOW IN ACTION**
- **FIRST MILE OF SALES**
- **FINAL WORD ON LEAD CAPTURE**

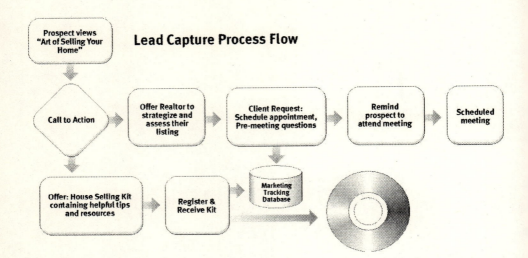

Lead Capture

"Do not trust your memory; it is a net full of holes; the most beautiful prizes slip through it."

-Georges Duhamel
The Heart's Domain

"They may forget what you said, but they will never forget how you made them feel."

-Carl W. Buechner

"Eighty percent of success is showing up."

-Woody Allen

The Last Mile of Marketing (or First Mile of Sales)

Now that you have launched a successful education-based marketing campaign (or EBM), your prospects begin to look to you for advice. You are becoming the trusted advisor. How do you convert trust into leads and sales? This chapter shows how to convert trust into leads at the prospects' own dictated pace, and transform these leads into qualified prospects for sales.

The goal of lead capture is to help the prospect gain even more information and be in control of the sales process. You are more likely to close a sale if the prospect is the one driving the buying process.

The Traditional Problem

Ultimately the goal of every marketing campaign is to create interested prospects, or leads, that end up becoming customers. Marketing has always been challenged with getting such leads to the sales department in a way that sales can process them and create new customers. Sales is typically too busy with prospects that they are already working with. Somewhere between marketing and sales, balls drop. Prospects who need attention receive poor service and consider alternative products and solutions. We call this the last mile of marketing.

When executed correctly, you should finish this last mile with well-qualified prospects who are seriously considering becoming customers of the service or products that you provide. Execute poorly, and you'll have educated a qualified prospect who moves to a competitor for a better customer experience.

The EBM Solution: Turn questions into sales leads

EBM campaigns provide a different opportunity for lead capture. Your prospects have a higher level of understanding about their problems and potential solutions. They are also open to hearing more about how you can solve their problems because they have already received value from your EBM campaign. They are looking to you as their trusted advisor.

Lead capture in this context is about using calls to action that are helpful in solving problems or providing further information to the prospect. Most people want to buy products and services when they are ready. They do not want to be sold to. EBM increases and accelerates your prospects' readiness for your products. At this stage, you need to provide ways to channel prospects' interest and questions into your sales process. Some helpful calls to action would be:

- Provide further information on how your products can solve the problems discussed in the EBM program
- Offer to assess their needs
- Offer to set up appointments with a company expert to help them visualize solutions.

EBM Lead Capture Benefits

Lead capture the EBM way has these benefits:

- Customers become active participants in the sales process. They are engaged as their questions are being answered.
- You learn more about your customers' needs and challenges.

- If you answer their questions well, you earn their allegiance and business, even if you might be charging a higher price than your competitors.

To be effective, remember to provide various options for helpful advice and service. Allow customers to choose how they want to interact with you, and how they prefer to examine your products and service.

EBM Differentiator

The traditional prospecting approach is to drown your prospect with product information and benefit statements that demonstrate why your offering is superior to other offerings in the marketplace. EBM lead capture, on the other hand, is about continuing to service the customer by providing more answers to questions, building upon the initial trust that was created through the EBM campaign. You guide the prospect into your sales process through a consultative and caring approach.

EBM lead capture is different from your typical sales call to action. Instead of hard selling your products' features and benefits, continue your role as the helpful advisor. People have a desire to be helped. They want to learn more about their options and be empowered to make the right decision. Lead capture here is about fulfilling needs and providing resources to empower your prospect to buy.

With this in mind, it is very important to create needs-based calls to action. Offer additional assistance, answer questions, and set up consultative assessment appointments for your clients when they request it.

Lead capture process flow

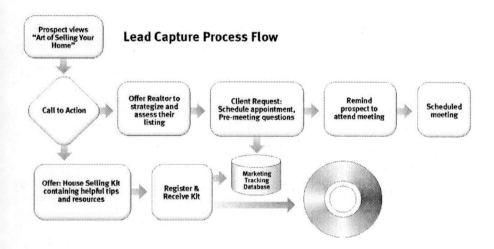

Prospect watches EBM show

Let's use "The Art of Selling Your Home" as an example of how lead capture would work. The prospect has watched the show, and her interest has been piqued. For this program, we might have two helpful calls to action:

- Offer to have a REALTOR help assess her property value and potential listing.

Lead Capture

- Offer a House Selling Kit complete with helpful tips and checklists to guide her through the complex process of listing and selling her home.

Process the lead

Now that you have a prospect interested and even responding to your call to action, you must take advantage of this achievement and provide excellent customer service as the prospect enters your sales process. It is amazing how many companies consistently do poorly in capturing a lead and guiding the prospect through the buying process. Here is where most marketing departments hand off prospects to sales, expecting the sales department to magically take over.

To guide the prospect through the last mile of marketing and first mile of sales, begin by capturing the following information:

- **Potential meeting times**: Ask the prospect for a few meeting dates that are convenient. Work with your sales team to schedule an appointment for this prospect. Do not leave it solely to the sales team as they do not have any relationship with the prospect at this point.

- **Pre-meeting questions**: Ask the prospect two or three questions that will enable your sales team to plan for a much more productive meeting. This is a benefit to the prospect as she will experience a smooth and knowledgeable transition. Use this information to create a prospect profile for sales. This profile should also include information on what content she viewed or documents she downloaded.

Confirm Appointment

Finally, don't forget to confirm the appointment with the prospect reminding her either by phone or email.

First mile of sales

Qualifying the Prospect

Qualifying the prospect is the first mile of sales. Prospects that arrive through EBM campaigns are already better qualified than typical prospects. They have already experienced value from you through your EBM content. They also have already identified pain and are looking actively for a solution to their pain.

What if they are not qualified?

However, there may still be prospects coming through who are not qualified. For example, they may have a need, but their budgets may not be in line with the typical projects you take on. Another example might be that they are looking for information to educate and prepare themselves for an upcoming buying season. In both of these cases, the sales manager needs to decide if the prospect is a good fit based on their company's qualifying criteria.

If the person is not qualified, I would still provide advice and information, and gently let them know they are not ready for your offering. This will create good will, impress the prospect, and might lead to referrals. Remember, too, that a promising prospect might not be ready to buy now, but could be someday. So if timing is a problem, simply continue to provide information and helpful advice. Continue to nurture the prospect, always offering value. Allow the fruit to ripen naturally, and you will harvest sweet rewards when your prospects are ready.

Final word on lead capture

Remember that the customer is trying to solve a problem or challenge. The best sales method is to allow the customer to buy rather than to push a sale. Allow the customer to buy by providing helpful information and world-class service throughout your lead capture and sales process, and you will be rewarded with sales and word-of-mouth referrals.

I'll end with a story that demonstrates this helpful attitude. My wife was shopping at a Nordstrom in Phoenix while we were visiting friends. She found a dress that she liked. The Nordstrom sales person educated her on how this designer was unique, and what made this a great dress for her. Unfortunately, her size was not available that day in that store. The sales person who provided great information and education was then able to switch to world-class customer service. Her first step was to offer to bring the dress in from another store. We were leaving the next day, so that did not work. Her second step was to offer to send the dress to us. My wife really wanted to try it on before buying. No problem – we could send it back to her if we did not like it, with the postage paid by Nordstrom. Her third step completely blew us away. She asked if she could put us on her mailing list and continue to shop for us remotely. My wife could try on her selections, and send back anything she didn't like, with Nordstrom paying for all the shipping. Startled at this generous offer, we agreed.

Not only is this a stellar example of world-class education-based marketing and customer service, it also demonstrates how to capture a highly qualified lead. The salesperson gathered my wife's size and tastes, along with our address, phone numbers, and email. We provided all the information she requested because we trusted the sales person, and her calls to action were irresistible. As a result, my wife has become a Nordstrom customer for life.

Educating and showing concern for your customers earns their trust and heart. Marry that with world-class customer service and you earn a life-long customer.

Chapter Summary

Bottom Line

It is better to have fewer qualified leads rather than many more unqualified leads. Use your program's "Calls To Action" to allow prospects to self qualify. Offer to answer questions or provide additional helpful information and continue to allow the prospect to make a better educated buying decision.

How?

Allow the customer to buy rather than selling to them. Marry education-based marketing with world-class customer service. Customers will naturally want to provide their contact information to you if your information has truly been helpful and valuable to them.

Thought Starters

1. What questions would you like to ask prospects coming to your site to qualify them for business?

2. What are the two most valuable questions you can ask your customers if they would answer them for you? Why are these the most valuable questions?

Now It's Your Turn

In the course of this book, we've described the three dramatic changes affecting consumers:

- Customers have more information.

- They have more choices.

- They are overloaded with information. So many people are sending out so many marketing messages, consumers just tune them out.

You can cut through the clutter by building trust. You do so with education-based marketing. You've learned our step-by-step approach to capture what your business knows and package that knowledge into stories. These stories will build your customers' and prospects' trust in you. When you are their trusted advisor, your business is in a much better position to grow. The opportunities are exciting. Take advantage of them!

Glossary

Education-based marketing - A marketing method that emphasizes sharing information and helpful advice to build credibility and trust. This style of marketing works best for "big, sweaty and passionate" subjects. Select a topic that involves at least two of these critical elements: expense, worry, and passion. Having all three is even better. Education-based marketing works best when your prospects already care passionately about your subject.

Encoding - The process of compressing video files into a significantly smaller format for the purpose of streaming over the Internet.

Expert - Your own company's expert on your topic. In a story package, the best role for the expert is to reinforce what the viewer learned from the hero.

Hero - A person who has done what you are trying to teach. Ideally, you will build your story package around a hero.

Host - The host serves as the "glue" to hold the story package together, voicing over the essence of the story and why you will want to watch. The host provides the teases, the story set up and any concluding or summary comments.

KPS - the rate of data transmission. 100kps is slower than 300kps. The higher the speed, the higher the video quality and the larger the bandwidth necessary for viewing the content.

Luminary - A third party or outside expert on your topic.

MediaPlayer - The software program that allows you to watch videos or listen to music on computer.

Segment - One or more story packages that fit together. When you have a lot of content to communicate, divide it into segments. No segment should be longer than five minutes.

Segment map - A high-level view of what will be covered in each segment, including topic, title, tease, hero story, luminary, expert and transferrable lesson.

Story package - A term borrowed from television. It usually implies a reporter has interviewed several people, added a voiceover, and edited the footage into a "package." In our case, you and your colleagues will be the reporter. Story packages are usually less than two minutes long.

Tease - An enticement that hooks the viewer to stay and watch your content. It should grab your viewer in the first 10 to 15 seconds. It answers the question, "Why should we watch this?"

Transferrable Lesson - A list of the key things your audience should know about your topic. The transferrable lesson is the final part of your story.

Index

Symbols

3M 52, 53, 57, 59
5-Step process 35, 37

A

Audience 61

B

Bold Comparison 52, 54
Business case
   ROI 63

C

Call to action 62
Category 52, 53
CD-ROM 6, 29, 83, 84, 85, 88, 95, 100, 101, 112
Companies mentioned
  3M 52, 53, 57, 59
  BMW 14
  Georgia-Pacific 84, 85
  LendingTree 6
  Wachovia 6, 91, 94, 95, 96, 99, 100, 101
  Yahoo! 6, 7, 8
Consumer 14, 15, 16, 26, 84, 114
Content 6, 67, 69, 83, 89, 120
  categories 53, 59
  executive buy in 61
  lists 60, 61, 73
  the body 49, 51, 52
Content Boot Camp 67, 69
Customer service 70

D

Delivery channels
  CD-ROM 6, 29, 83, 84, 85, 88, 95, 100, 101, 112
  DVD 29, 113, 114, 115
  VHS 112, 114
Delivery Platforms
  Apple Macintosh 118
  Microsoft 118
  QuickTime 117
  RealPlayer 117
  Windows Media Player 117
Design 67, 69, 79, 84
  Content Boot Camp 67, 69
  Segment map 67, 77, 138
Develop 65, 81, 83
DVD 29, 113, 114, 115

E

EBM 24, 31, 63, 123, 125, 126, 127, 128, 129, 131
  Education-Based Marketing 21, 23, 61, 67, 71
Encode
  Encoding 117, 137
Expert 72, 137

H

Hero 72, 137
Host 137

I

Internet 5, 6, 14, 28, 84, 86, 88, 89, 98, 100, 101, 110, 112, 116, 117, 137
  broadband 117

K

Knowledge
  capture knowledge 37

Index

L

Laws
  law of hooks, teases, titles 72
  law of metaphors 73
  law of the story 71
  law of three perspectives 72
Lead capture 123, 127, 128, 129
  Benefits 127
  confirm appointment 131
  Last mile of marketing 125
  process the lead 130
  qualifying 131
Luminary 72, 138

M

Marketing
  Ads 15
  closing the loop 102, 107, 118
  Direct mail 96
  Monitor and refine 94, 102
  Metrics 119, 120
    Call To Action 120
    Clicked to Buy 121
    Conversion to Customers 121
    Conversion to Viewers 120
    Marketing Response 120
MindBlazer 5, 6, 7, 8, 44, 70, 94
Multi-channel 91, 94, 99

O

Observe 41, 43, 47
Offline 112
Online 107, 112, 115

P

Post-it 52, 53, 57, 59
Problem/Solution 52, 56
Production 81, 86, 97

Q

QA
  QA and review 91, 96

R

Reach 28
Review 94, 97, 98, 120
Rollout 91, 93, 103

S

Segment 62, 67, 77, 138
  map 67, 77, 138
Segments 62
Sequential 52
Story 7, 57, 71, 107, 109, 138
  package 7, 138
Story packages 7, 138
Streaming 115

T

Tease 49, 51, 58, 138
Technology 70, 84, 105, 107, 109, 122
Training 93
Transferrable Lesson 49, 51, 60, 72, 97, 138
TV 5, 6, 7, 13, 15, 29, 58, 63, 73, 75, 86, 94, 95, 96, 97, 101
  TV world 5, 7, 58

V

VHS 112, 114
Viewer 71, 115
Viewers 62, 73, 75, 116, 120
Visualize 49, 51, 65

W

webcast 6, 8

Y

Yahoo! 6, 7, 8

Printed in the United States
19516LVS00004B/130-132